PUBLISHER COMMENTARY

We print NASA's handbooks and standards for the convenience of those that use them on a daily basis. We print all of these a full 8 ½ by 11 with large text so they are easy to read. Yes, color books are expensive to print so unless the information relies on the use of color for proper interpretation or understanding, we print most books in black and white to keep the cost down. All these documents are available for download for free from NASA, however printing them all over a network printer would take days.

Why buy a book you can download free? We print this so you don't have to.

All these books are available for free download from the government web site. Some are available only in electronic media. Some online docs are missing pages or barely legible.

We at 4th Watch Publishing are former government employees, so we know how government employees actually use the standards. When a new standard is released, an engineer prints it out, punches holes and puts it in a 3-ring binder. While this is not a big deal for a 5 or 10-page document, many NIST documents are over 100 pages and printing a large document is a time-consuming effort. So, an engineer that's paid $75 an hour is spending hours simply printing out the tools needed to do the job. That's time that could be better spent doing engineering. We publish these documents so engineers can focus on what they were hired to do – engineering. It's much more cost-effective to just order the latest version from Amazon.com

If there is a standard you would like published, let us know. Our web site is www.usgovpub.com

www.usgovpub.com

NASA TECHNICAL STANDARD

National Aeronautics and Space Administration

**NASA-STD-5006A
w/CHANGE 1:
REVALIDATED
w/ADMINISTRATIVE/
EDITORIAL CHANGES
2016-05-03**

Approved: 2015-07-31
Superseding NASA-STD-5006

GENERAL WELDING REQUIREMENTS FOR AEROSPACE MATERIALS

DOCUMENT HISTORY LOG

Status	Document Revision	Change Number	Approval Date	Description
Baseline			1999-02-17	Baseline Release
Revision	A		2015-07-31	General Revision
		1	2016-05-03	Revalidated w/Administrative/Editorial Changes—This NASA Technical Standard was reviewed and no technical changes resulted. Administrative changes to number requirements, add a Requirements Compliance Matrix as an appendix, and conform to the current template were made, along with editorial corrections. ASTM E2375 was added as a reference document.

FOREWORD

This NASA Technical Standard is published by the National Aeronautics and Space Administration (NASA) to provide uniform engineering and technical requirements for processes, procedures, practices, and methods that have been endorsed as standard for NASA programs and projects, including requirements for selection, application, and design criteria of an item.

This NASA Technical Standard is approved for use by NASA Headquarters and NASA Centers and Facilities and may be cited in contract, program, and other Agency documents as a technical requirement. It may also apply to the Jet Propulsion Laboratory and other contractors only to the extent specified or referenced in applicable contracts.

This NASA Technical Standard establishes general directions and describes the type of information that NASA expects for welded structures. This NASA Technical Standard does not provide the detailed process and quality assurance requirements for weldments on flight hardware. Instead, it is intended as a higher level document which states minimum requirements for welded hardware.

Requests for information should be submitted via "Feedback" at https://standards.nasa.gov. Requests for changes to this NASA Technical Standard should be submitted via MSFC Form 4657, Change Request for a NASA Engineering Standard.

Original Signed By: 07/31/2015

_____   _____
Ralph R. Roe, Jr.   Approval Date
NASA Chief Engineer

TABLE OF CONTENTS

LIST OF APPENDICES

LIST OF FIGURES

GENERAL WELDING REQUIREMENTS FOR AEROSPACE MATERIALS

1. SCOPE

1.1 Purpose

The purpose of this NASA Technical Standard, as defined in NASA Procedural Requirement (NPR) 7120.10, Technical Standards for NASA Programs and Projects, is to establish the processing and quality assurance requirements for manual, automatic, machine, and semiautomatic welding for spaceflight applications and special test equipment used for testing flight hardware with the exception of ground-based pressure systems, which are subject to NASA-STD-8719.17, NASA Requirements for Ground-Based Pressure Vessels and Pressurized Systems (PVS).

1.2 Applicability

This NASA Technical Standard is applicable to all welding processes used for joining metallic materials. This includes, but is not limited to arc welding (AW), solid state welding (SSW), resistance welding (RW), and high energy density welding (HEDW). This NASA Technical Standard covers all metallic materials used in the manufacture of hardware for spaceflight applications and special test equipment used for testing flight within NASA.

This NASA Technical Standard is approved for use by NASA Headquarters and NASA Centers and Facilities and may be cited in contract, program, and other Agency documents as a technical requirement. It may also apply to the Jet Propulsion Laboratory and other contractors only to the extent specified or referenced in applicable contracts.

Verifiable requirement statements are numbered and indicated by the word "shall"; this NASA Technical Standard contains 142 requirements. Explanatory or guidance text is indicated in italics beginning in section 4. To facilitate requirements selection and verification by NASA programs and projects, a Requirements Compliance Matrix is provided in Appendix B.

1.3 Tailoring

[GWR 1] Tailoring of this NASA Technical Standard for application to a specific program or project shall be formally documented as part of program or project requirements and approved by the responsible Technical Authority in accordance with NPR 7120.5, NASA Space Flight Program and Project Management Requirements. The requirements in this document may be tailored by submitting a detailed weld process specification or stating applicable industry standards that meet the intent of this NASA Technical Standard.

2. APPLICABLE DOCUMENTS

2.1 General

The documents listed in this section contain provisions that constitute requirements of this NASA Technical Standard as cited in the text.

2.1.1 [GWR 2] The latest issuances of cited documents shall apply unless specific versions are designated.

2.1.2 [GWR 3] Non-use of specifically designated versions shall be approved by the responsible Technical Authority.

The applicable documents are accessible at https://standards.nasa.gov, may be obtained directly from the Standards Developing Body or other document distributors, or information for obtaining the document is provided.

2.2 Government Documents

National Aeronautics and Space Administration (NASA)

NPR 1441.1	NASA Records Management Program Requirements
NPR 7120.5	NASA Space Flight Program and Project Management Requirements
NPR 7120.10	Technical Standards for NASA Programs and Projects

2.3 Non-Government Documents

Aerospace Industries Association (AIA)/National Aerospace Standard (NAS)

| AIA/NAS 410 | NAS Certification and Qualification of Nondestructive Test Personnel |

American Welding Society (AWS)

| AWS QC1 | Standard for AWS Certification of Welding Inspectors |

2.4 Order of Precedence

2.4.1 The requirements and standard practices established in this NASA Technical Standard do not supersede or waive existing requirements and standard practices found in other Agency documentation.

2.4.2 [GWR 4] Conflicts between this NASA Technical Standard and other requirements documents shall be resolved by the responsible Technical Authority.

3.   ACRONYMS AND DEFINITIONS

3.1   Acronyms and Abbreviations

AIA	Aerospace Industries Association
AMS	Aerospace Material Specification
ASME	American Society of Mechanical Engineers
ASTM	American Society for Testing Materials
AW	arc welding
AWS	American Welding Society
FSW	friction stir welding
GWR	general welding requirement
HEDW	high energy density welding
JSC	Johnson Space Center
MIL	military
MPR	Marshall Procedural Requirements
MRB	Material Review Board
MSFC	Marshall Space Flight Center
NAS	National Aerospace Standard
NASA	National Aeronautics and Space Administration
NDE	nondestructive evaluation
NPR	NASA Procedural Requirements
PQR	Procedure Qualification Record
PVS	pressure vessels and pressurized systems
QA	Quality Assurance
QC1	Standard for AWS Certification of Welding Inspectors
RW	resistance welding
S&MA	Safety and Mission Assurance
SAE	Society of Automotive Engineers
SSW	solid state welding
STD	standard
WPS	Welding Procedure Specification

3.2   Definitions

Automatic Welding: A welding operation performed without adjustment of the controls by a welding operator.

Certified: With respect to a welder, means a welder or inspector who has passed qualification tests based on requirements established in this NASA Technical Standard. With respect to a procedure or process specification, a term describing a weld procedure or process that has passed qualification tests based on requirements established in this NASA Technical Standard.

Concave Root Surface: A weld root with penetration not extending beyond the thickness of the parent metal. *Note: Periodically referred to as "suckback."*

Conventional Friction Stir Welding (FSW): FSW in which the load is reacted by an anvil.

Critical Flaw Size: The analytically determined flaw size that produces a critical stress intensity factor of concern for a specified number of life cycles which will likely produce a catastrophic mission failure.

Defect: A discontinuity or discontinuities that by nature, or accumulated effect, render a part or product unable to meet minimum standards or specifications; designates rejectability.

Dross: A mass of solid impurities floating on a molten metal or dispersed in the metal.

Essential Variables: Weld process parameters that influence directly the weld process and resulting weld properties in such a manner that changes to them require requalification of the weld procedure. *Note: Examples are heat input, travel speed, torch setup, and pin tool configuration.*

Fail Safe: A condition in which, after failure of a single individual structural member, the remaining structure can withstand the redistributed loads with an ultimate factor of safety of 1.0 on limit load. *Note: The failure is contained or constrained so that the failed part does not affect other flight elements or personnel.*

Heat-Affected Zone: The portion of the base metal whose microstructure or mechanical properties have been altered by the heat of welding, brazing, soldering, or thermal cutting.

Heat Input: Quantity of energy introduced per unit length of weld from a traveling heat source, expressed in joules per millimeter or joules per inch. *Note: Computed as the ratio of the total input power of the heat source in watts to the travel velocity in millimeters per second or inches per minute.*

Heat Sensitive Alloys: Alloys that require mechanical working, precipitation strengthening, or other metallurgical mechanisms to regain their rated strength due to exposure to the heat input from the welding process.

Improper Fusion: A condition when the weld metal that replaces base metal is insufficient.

Incomplete Fusion: A weld discontinuity in which fusion did not occur between weld metal and parent material or adjoining weld beads.

Incomplete Joint Penetration: The condition of a weld failing to extend through the full thickness of the joint.

In-Process Correction: Action taken by a welder to complete a process before submittal to inspection.

Lack of Fill: A weld face surface not extending to the surface of the parent metal.

Machine Welding: Welding with equipment that performs the welding operation under the constant observation and control of a welding operator.

Manual Welding: A welding operation performed and controlled completely by hand.

Material Review Board (MRB): A cross-functional group that reviews production or purchased items on hold because of nonconformance or usability concerns. The MRB is to determine the disposition, which may include rework, scrap, or return to the vendor.

Material Thickness: The minimum material thickness of a joint member per drawing tolerance. *Note: The thinner of the joint members with different thicknesses is designated "t."*

Mismatch: The linear misalignment of components resulting from improper fit-up or distortion during welding. *Note: Mismatch is calculated as the difference in the alignment of specified features (usually either center lines or surfaces) of the two parts having been welded and should not be confused with the difference in center lines as a result of welding two different thickness components.*

Nonstructural Weld: A non-load-bearing weld.

Peaking: The angular distortion of the components resulting from welding. *Note: Peaking is calculated as the angle resulting from the intersection of tangents taken from the surface of the two components being welded.*

Procedure Qualification Record (PQR): A document providing the actual welding variables used to produce an acceptable test weld and the results of tests conducted on the weld for the purpose of demonstrating process and procedural capability and repeatability. *Note: Demonstration of capability qualifies the welding procedure.*

Qualified Inspector: A certified individual with the responsibility and ability to judge the quality of the welded specimens in relation to some form of written specification.

Repair: A procedure that makes a nonconforming item acceptable for use. *Note: The purpose of the repair is to reduce the effect of the nonconformance. Repair is distinguished from rework in that the characteristics after repair still do not completely conform to the applicable drawings, specifications, or contract requirements. Nonstandard repair procedures are authorized by MRB action for use on a one-time basis only. All repairs require MRB approval before implementation.*

Rework: A procedure applied to a nonconforming item that completely eliminates the nonconformance and results in a characteristic that conforms completely to the drawings, specifications, or contract requirements. *Note: Not all rework activities require MRB approval before implementation.*

Semiautomatic Welding: Welding with equipment that controls only the filler metal feed. *Note: The weld progression is manually controlled.*

Special Test Equipment: Any non-flight, non-GSE, or non-facility structure, hardware, piping systems, pressure vessels, or equipment intended to be used for testing or simulation, or associated with the manufacturing, process development, and preparation of Marshall Space Flight Center (MSFC) facilities for testing or simulation. *Note: Designs include, but are not limited to, test stands, test beds, load reaction and application structures, load line components, hot fire testing of engines and engine components, fluid flow and pressure tests, high pressure and/or cryogenic storage and/or run systems, solid propellant tests, flight hardware mockups and simulators, hardware support stands and dollies, personnel access stands, lifting and handling hardware, and tooling used to facilitate the fabrication and/or assembly of flight/non-flight hardware, such as master drill templates or alignment/clamping fixtures used during machining and welding processes.*

Undercut: A groove melted into the base metal adjacent to the weld toe or weld root and left unfilled by weld metal.

Welding Procedure Specification (WPS): A document providing in detail the required variables for a specific application to ensure repeatability by properly trained welders and welding operators.

Welding Process Specification: A document that prescribes, in a complete, precise, verifiable manner, the requirements, design, behavior, or characteristics of a system or system component.

Weld Zone: The weld metal fusion zone plus the heat-affected zone.

4. REQUIREMENTS

4.1 Specification of this NASA Technical Standard on Contracts

[GWR 5] When this NASA Technical Standard is specified on contract documents, a detailed weld process specification, as defined in NPR 7120.10, which meets the intent of this NASA Technical Standard shall be submitted. *Industry, government, and company specifications can be used for welding flight hardware if they contain the information required by this NASA Technical Standard. The contractor has the responsibility to submit the detailed weld process specification.*

4.2 Joint Classes

[GWR 6] Welding performed using this NASA Technical Standard shall be classified in accordance with the service of the joints as follows in the next sections.

4.2.1 Class A

Critical applications. Welds where a single failure would cause loss of system, loss of major components, loss of control, and loss of crew.

4.2.1.1 [GWR 7] Class A welds shall require visual, dimensional, surface, and volumetric inspections, and additional inspection when required by engineering drawing.

Note: Based on consequences of failure, all fracture-critical welds are, by definition, Class A joint. If the quality of the Class A joint cannot be verified as required by this NASA Technical Standard, e.g., inaccessible volume or root surfaces, then alternative rationale for acceptance is to be presented to the responsible NASA Fracture Control Board for approval as required by NASA-STD 5019, Fracture Control Requirements for Spaceflight Hardware.

4.2.2 Class B

Semicritical applications. Welds where a failure would reduce overall efficiency of the system, preclude the intended function or use of the equipment, but loss of the system or endangering personnel would not be experienced.

4.2.2.1 [GWR 8] Class B welds shall require visual, dimensional, and surface inspections, and additional inspection when required by engineering drawing.

4.2.2.2 [GWR 9] Class B welds shall be subjected to volumetric inspection if required by engineering design and specified by drawing or special instruction.

4.2.2.3 [GWR 10] Weld requiring fail-safe capability shall be classified as a Class B joint.

4.2.3 Class C

Noncritical applications. Welds where a failure would not affect the efficiency of the system or endanger personnel.

4.2.3.1 [GWR 11] Class C welds shall require visual and dimensional inspections, and additional inspection when required by engineering drawing.

4.2.3.2 [GWR 12] Class C joints shall require weld integrity verification based on function of the joint (e.g., seal welds require leak testing commensurate with the leak rate requirement).

4.3 Equipment

4.3.1 Welding Equipment

a. [GWR 13] Automatic, semiautomatic, manual, and machine welding shall be accomplished using equipment containing calibrated dials, meters, or recorders that quantitatively indicate process parameters.

b. [GWR 14] All joining equipment (including manual) shall be capable of producing joints that meet the requirements specified herein.

4.3.1.1 Acceptance Testing

a. [GWR 15] New, repaired, relocated, or modified welding machines and equipment for automatic and machine welding shall be acceptance-tested prior to processing of flight hardware.

b. [GWR 16] Machines and equipment shall meet the requirements of the applicable purchase specification, design specification, or modification order.

c. [GWR 17] Power supplies and supporting components (electrical or mechanical or both) shall be capable of operating reliably within the range of parameters and duty cycle to be used for joining production parts.

4.3.1.2 Calibration

a. [GWR 18] Measuring instruments, meters, gages, or direct reading electrical control circuits to be used for automatic, semiautomatic, and machine joining operations shall be initially calibrated and periodically recalibrated to maintain adequate performance.

b. [GWR 19] Maintenance performed on measuring instruments, meters, gages, or direct reading electrical control circuits to be used for automatic, semiautomatic, and machine performance joining operations shall require recalibration to maintain adequate performance.

c. [GWR 20] Measuring instruments, meters, gages, or direct reading electrical control circuits to be used for automatic, semiautomatic, and machine joining operations shall be initially calibrated and periodically recalibrated to maintain adequate performance or when any maintenance is performed that may have changed calibration.

4.3.1.3 Maintenance and Records

a. [GWR 21] Welding machines shall be provided with adequate periodic maintenance service so that acceptable welds can be produced using qualified welding procedure specifications.

b. [GWR 22] A current record of each maintenance repair or functional check shall be maintained for each welding machine.

4.3.2 Tooling and Fixtures

a. [GWR 23] Tooling and fixtures used in the joining operation shall be constructed of nonmagnetic materials that do not affect the welding arc or beam, or that are not detrimental to the weld quality.

b. [GWR 24] Tooling and fixtures shall not be a source of contamination of the joint.

c. [GWR 25] Magnetic materials, when used, shall be degaussed prior to welding.

d. [GWR 26] Degaussing, when necessary, shall be controlled by the WPS for the successful completion of the weld.

e. [GWR 27] Tooling and fixtures required to ensure compliance with dimensional requirements of section 4.8.3 shall be identified on the WPS.

4.4 Materials

4.4.1 Base Metals

a. [GWR 28] Unless otherwise specified or approved by the procuring agency, base metal alloy shall conform to applicable government and/or industry specifications for each alloy group.

b. [GWR 29] The base metal, material condition, and appropriate specification shall be recorded in the WPS.

c. [GWR 30] Weld start and runoff tabs, when used, shall be of the same alloy as the material being joined and be welded with the same filler metal specified on the drawing or WPS. *Backing material may be used when verified by procedure qualification.*

4.4.2 Filler Metals

a. [GWR 31] Unless otherwise specified or approved by the procuring agency, filler metal alloy shall conform to applicable government and/or industry specifications for each alloy group.

b. [GWR 32] Specifications used to procure filler metals shall include provisions to mitigate the possibility of two different filler wires being errantly mixed together on a single spool or in a filler rod container.

c. [GWR 33] Weld filler metals and the appropriate specifications shall be recorded in the WPS.

4.4.3 Shielding Gas

a. [GWR 34] Welding-grade gases conforming to the applicable industry or military specifications shall be used for gas shielding.

b. [GWR 35] The shield gas type, specification, and flow rate shall be recorded in the WPS.

4.4.4 Tungsten Electrodes

a. [GWR 36] Tungsten electrodes shall conform to the applicable industry or military specifications.

b. [GWR 37] The electrode diameter, electrode tip shape, alloy composition, and specification shall be recorded as a part of the WPS.

4.4.5 Friction Stir Welding Pin Tools

4.4.5.1 [GWR 38] Pin and shoulder service life shall be demonstrated to meet the intended use.

4.4.5.2 [GWR 39] Pins and shoulders shall be limited to the demonstrated life.

4.4.5.3 [GWR 40] Pin tool design, materials, and service life shall be recorded in the WPS.

a. [GWR 41] Pins and shoulders that have reached the specified service life shall be marked and removed from service to preclude an accidental future use in the FSW production process.

b. [GWR 42] If used for more than one weld joint, pins and shoulders shall be cleaned and inspected before reuse on production hardware.

APPROVED FOR PUBLIC RELEASE — DISTRIBUTION IS UNLIMITED

4.5   Weld Procedure and Performance Qualification

4.5.1   Welder Performance Qualification

a.   [GWR 43] Operators of automatic, semiautomatic, machine, or manual welding equipment shall be certified for the applicable process.

b.   [GWR 44] Suppliers shall define the weld certification process in their detailed weld process specification.

4.5.2   Welding Procedure Specification

a.   [GWR 45] Prior to first production of parts, or when changes are made to essential variables of the WPS, qualification joints shall be made to establish a satisfactory WPS for each different configuration of A, B, and C classes of welds. *Figure 1, Welding Procedure Specification Example, contains an example of a WPS (for reference only).*

b.   [GWR 46] Variables considered essential shall be so identified in the WPS.

4.5.2.1   Classes A and B Joints

a.   [GWR 47] Classes A and B joints shall be qualified with joints that simulate the production part with respect to section thickness, alloy, heat-treat condition, joint preparation, pre-weld cleaning, fit-up, position, and post-weld operations.

b.   [GWR 48] The joints shall be processed in either the actual production fixture or in a test fixture simulating the production fixture using the production welding equipment.

c.   [GWR 49] Base metal for qualification joining tests shall be identified by lot or heat number, type, and condition and maintain identification through all evaluation processes.

d.   [GWR 50] The qualification weld shall be subjected to metallurgical evaluation and the same post-weld inspections and processes as the production parts, including reinforcement removal, mechanical deformation, stress relief, and thermal treatments associated with artificial aging or any operation affecting mechanical properties.

Weld Panel ID	Program Code	Material Type	Material Heat Lot Number	Material Manufacturer	Material Thickness	Material Ser. Num	Welding Process (dc-,dc+,vppa)

Operator	Electrode Type	Welding Torch	Welding Torch Shield Cup	Torch Orientation (Lead, Lag)	Welding Position (Vertical)	Back Purge Gas Type	Plasma Gas Type	Shield Gas Type	Trailing Shield Gas Type

Building Number	Power Supply	Weld Fixture	Weld Station	Back Purge Type	Filler Wire Type	Filler Wire Heat Lot Number	Filler Wire Manufacturer	Trailing Shield Type

Electrode Configuration	Joint Configuration

Weld Passes	Welding Current	Welding Voltage	Shield Gas Flow (SCFH)	Plasma Gas Flow Rate	Plasma Gas Pressure	Filler Wire Size Dia.	Filler Wire Rate (IPM)	Travel Rate (IPM)	Interpass Temperature	Orifice Size
Tack Pass										
First Pass										
2nd Pass										
3rd Pass										
4th Pass										
5th Pass										
6th Pass										
7th Pass										
8th Pass										

Weld Passes	Electrode Size	Electrode Set Back	Back Purge Gas Flow	Back Purge Gas Pressure	Trailing Shield Gas Flow	Straight Polarity Time (ms)	Reverse Polarity Time (ms)	Added Reverse Current	Arc Oscillation Dwell	Arc Oscillation Frequency	Arc Oscillation Position	Arc Oscillation Amplitude
Tack Pass												
First Pass												
2nd Pass												
3rd Pass												
4th Pass												
5th Pass												
6th Pass												
7th Pass												
8th Pass												

Comments:

Figure 1—Welding Procedure Specification Example

4.5.2.2   Joining Parameters

a.   [GWR 51] As a minimum, all essential joining variables (such as voltage, current, rate of travel, position, and filler-wire feed rate) shall be recorded during qualification welding.

b.   [GWR 52] Manual weld parameters and operating parameter ranges shall be established during the WPS qualification.

c.   [GWR 53] The WPS shall document all pre-welding operations, setup conditions, welding equipment, and any pertinent information about the welding system used which affects the joining operation.

4.5.2.3   Parameter Tolerances

a.   [GWR 54] For automatic, semiautomatic, and machine joining, parameter tolerances may be used and shall be listed in the qualified WPS.

b.   [GWR 55] Test samples representing the minimum and maximum heat input shall be processed to verify acceptable welds and the results documented in the PQR.

4.5.3   Welding Procedure Specification Qualification

[GWR 56] All test and inspection results used to verify the weld integrity shall be recorded on the PQR.

4.5.4   Records

4.5.4.1   [GWR 57] Records of test specimens that meet the acceptance requirements of this process specification shall be signed and dated by a Quality Assurance (QA) representative as an accurate record of the welding and testing of the procedure test weldment.

4.5.4.2   [GWR 58] The WPS and PQR shall be prepared and retained as long-term temporary records in accordance with NPR 1441.1, NASA Records Retention Schedules, with the current WPS being maintained at the welding station.

4.5.4.3   [GWR 59] All WPSs and PQRs shall be maintained and made available for review by the responsible NASA Engineering Authority before production of hardware covered under this NASA Technical Standard.

4.6   Pre-Weld Operations

4.6.1   Joint Design

[GWR 60] All joints shall be documented on a WPS, design drawing, or other suitable document. *Acceptable joint designs are butt, lap, corner, tee, and edge.*

4.6.2   Pre-Weld Cleaning

a.   [GWR 61] Pre-weld cleaning of filler materials and surfaces to be welded to remove contaminants that are detrimental to weld quality shall be accomplished in an environment that will not degrade the quality of the weld.

b.   [GWR 62] Cleanliness shall be maintained during welding.

c.   [GWR 63] Pre-weld and interpass cleaning requirements shall be included in the WPS.

4.7   Production Welding

4.7.1   Equipment Operational Check

a.   [GWR 64] A welding equipment operational readiness check shall be made immediately prior to a production weld to verify the equipment is operating properly.

b.   [GWR 65] The equipment operational readiness check criteria shall be provided to the procuring agency.

4.7.2   Temperature Control

a.   [GWR 66] Pre-heat, interpass, and post-heat temperatures shall be controlled so as not to degrade the properties of the material being welded.

b.   [GWR 67] The parameters of pre-heat, interpass, and post-heat temperatures shall be recorded in an applicable WPS.

4.7.3   Tack Welding

a.   [GWR 68] Tack welding is allowable and shall either be removed or become a part of the finished weld (i.e., tack welds are to be completely consumed by the final weldment).

b.   [GWR 69] Tack welds that become part of the finished weld shall be performed by certified welders in accordance with certified procedures meeting the requirements of this NASA Technical Standard.

c.   [GWR 70] After the weldment is completed, the tack areas shall be evaluated to the requirements of the finished weld.

d.   [GWR 71] Tack welding requirements shall be included in the WPS.

4.7.4   Welding Techniques

4.7.4.1   Classes A and B Joints

a.   [GWR 72] The technique of welding the initial passes from both sides where the weld roots overlap beneath the exposed surfaces (reference figure 2 (A), Welding Techniques) shall be permitted only if the root of the first pass is removed to sound metal prior to placement of the first weld pass from the second side.

b.   [GWR 73] Joints which have prepared grooves from one or both sides (reference figure 2 (B) and (C)) and/or multi-pass welds shall have a weld land that is completely penetrated on the initial pass. *Partial penetration welds from one side are permissible provided the opposite side is machined into the penetration root prior to completing the weld.*

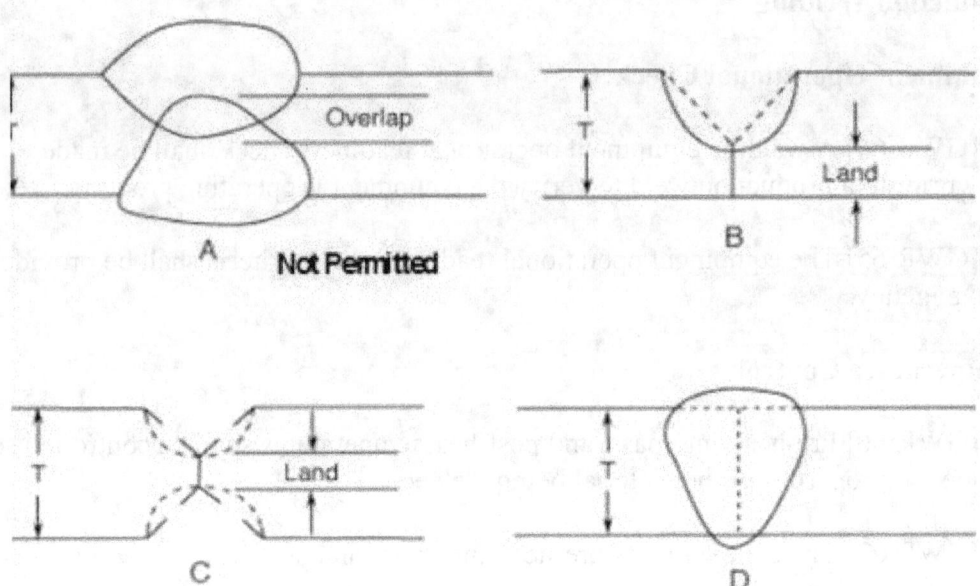

Figure 2—Welding Techniques

    c. [GWR 74] Adequate nondestructive evaluation (NDE) procedures shall be employed to ensure that the weld root has been exposed by machining.

    d. [GWR 75] All penetration weld passes shall have no visual evidence of improper fusion or presence of dross.

    e. [GWR 76] Square groove welds shall be completely penetrated from one side (reference figure 2 (D)).

4.7.4.2   Class C Joints

    a. [GWR 77] The technique of welding and joint geometry shall be as stated on the engineering drawing and the WPS.

    b. [GWR 78] Any deviation regarding weld technique and joint geometry shall be approved by the procuring agent prior to use.

    c. [GWR 79] Partial penetration groove welds shall be used only for Class C joints.

4.7.5 Welding Procedure

[GWR 80] Production welding shall be accomplished according to a qualified WPS. *A specific WPS for each weld is required for production welding Classes A, B, and C.*

4.7.6 Procedure Departure

a. [GWR 81] Departure from the qualified WPS during production welding shall require withholding the part for MRB disposition.

b. [GWR 82] The cause for departure shall be determined.

c. [GWR 83] Corrective action shall be taken prior to further production welding.

4.8 Post-Weld Operations

4.8.1 Inspection

[GWR 84] Each completed weldment, including the base metal, shall be inspected to ensure compliance with the requirements of sections 4.8.2, 4.8.3, and 4.10, and as dictated by the class of weld for a minimum of 12.5 mm (0.5 in) on either side of the weld interface.

4.8.2 General Visual/Surface Requirements

a. [GWR 85] Weld deposits, buildup, and root reinforcement shall comply with the criteria outlined in the accompanying detailed weld process specification that is submitted in support of this NASA Technical Standard.

b. [GWR 86] The edge of the weld deposit shall blend into the base metal without unfused overlaps or undercut.

c. [GWR 87] Weld face and root sides shall be free of surface cracks, crater cracks, and other defects open to the surface.

d. [GWR 88] Weld deposits shall be free of open voids or unfused overlapping folds or other lack of fusion.

e. [GWR 89] Undercutting, concavity, lack of fill, or root concavity shall be unacceptable in any weld where it occurs as a sharp notch or where the depth reduces the material thickness below the minimum thickness specified on the applicable drawing.

4.8.3 Dimensional Requirements

4.8.3.1 Mismatch

[GWR 90] If not specifically addressed in drawing tolerances or by specified welding standards, allowable post-weld mismatch shall be governed by overall drawing tolerances. *An example of weld joint mismatch is shown in figure 3, Mismatch and Peaking.*

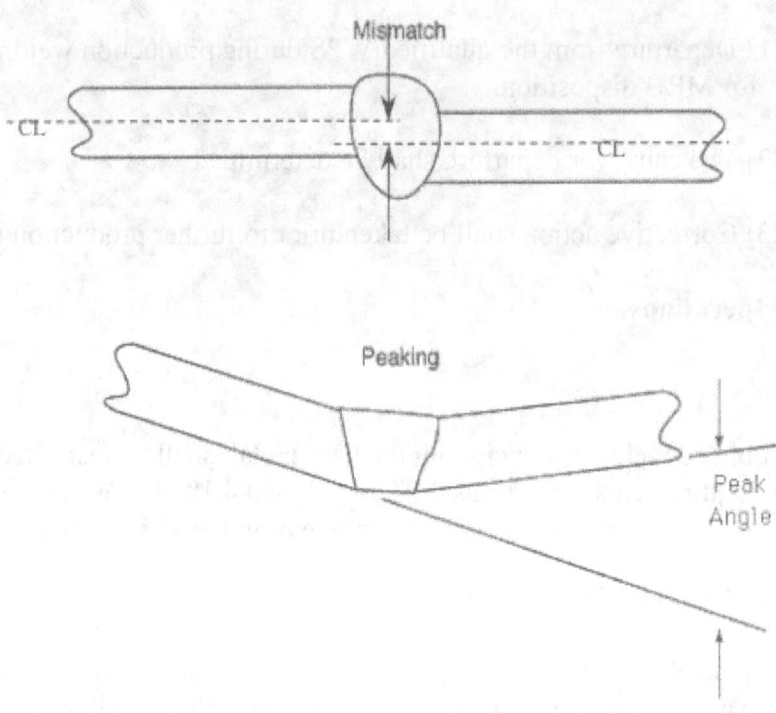

Figure 3—Mismatch and Peaking

4.8.3.2   Peaking

a. [GWR 91] If not specifically addressed in drawing tolerances or by specified welding standards, allowable post-weld peaking of the welded joint and adjacent base metal shall be governed by overall drawing tolerances.

b. [GWR 92] A standard template or other device having specified reference points shall be used for determination of peaking. *Weld peaking is shown in figure 3.*

4.8.3.3   Combination Mismatch and Peaking

The combined effect of mismatch and peaking on the efficiency of the weld joint is so related that one can be increased if the other is decreased. This condition can be tolerated if it can be shown by engineering analysis that positive margins of safety exist.

4.8.3.4   Weld Reinforcement Removal

a. [GWR 93] Weld bead reinforcement removal shall not thin the weld or parent metal below drawing dimensional requirements. *Weld bead reinforcement may be removed to eliminate defects occurring in the outer zones of the reinforcement unless otherwise specified on the engineering drawing.*

b.   [GWR 94] When flush contour is required by the welding symbol, weld reinforcement shall not exceed 0.4 mm (0.015 in).

c.   [GWR 95] Metal removal shall be such that the reworked area blends smoothly (e.g., 3.175 mm (0.125 in) radius) with adjacent material without abrupt sectional changes.

Surface grinding of base metal is permitted provided wall thickness is verified in compliance with dimensional requirements after grinding.

d.   [GWR 96] Weldments that are machined ground or otherwise mechanically worked causing disruption or smearing of the material surface shall be etched to remove the masking material before penetrant application.

4.8.3.5.1   Fillet Welds

a.   [GWR 97] Fillet weld fusion of the root shall have a minimum of 10 percent penetration of base metal thickness of the thinnest member of the root of the joint as determined by evaluation of transverse sections taken from the qualification welds.

(1)   [GWR 98] The minimum penetration shall be verified by destructive test/metallography.

(2)   [GWR 99] Weld parameters used to successfully and repeatedly complete the fillet welds shall be entered into the WPS and used for actual welding.

b.   [GWR 100] Fillet weld fusion of the root (reference figure 4, Fillet Welds) shall be determined by evaluation of transverse sections taken from the qualification welds.

c.   [GWR 101] Intermittent fillet welds shall have fusion of the root throughout the specified length. *Unless otherwise specified on the engineering drawing, the fillet may be extended by 6.35 mm (0.25 in) at each end without penetration in the extension.*

d.   [GWR 102] The minimum acceptable fillet size shall be that specified by the engineering drawing.

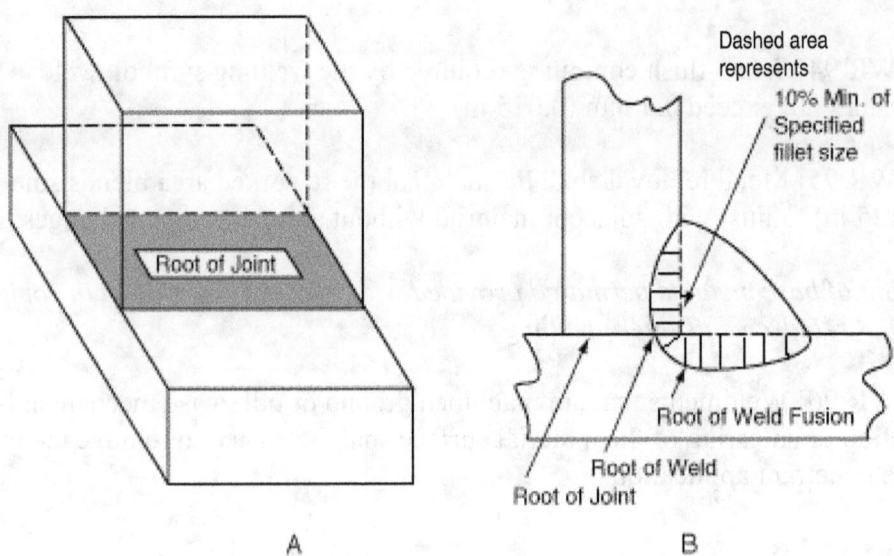

A   B

Notes: Root of Joint – That portion of a joint where members are
         closest to each other.

         Root of Weld – The point, as shown in cross section, at which
         the weld intersects the base metal surfaces.

The root of the weld shall penetrate to the extent that the actual throat
dimension exceeds the theoretical throat dimension, in addition, each
member shall be penetrated a minimum of 10% of the specified fillet
size at the root of the weld. Each leg length shall show fusion along the
surface of each common member.

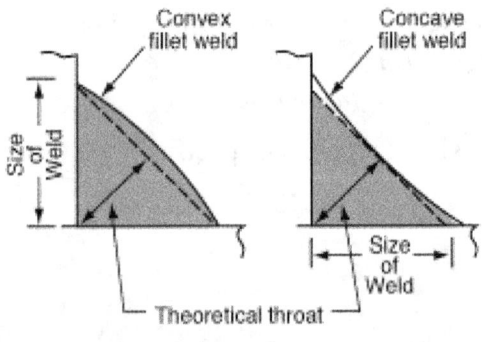

Equal leg fillet weld

For equal-leg fillet welds, the fillet size
is equal to the leg length of the largest
inscribed right isosceles triangle.

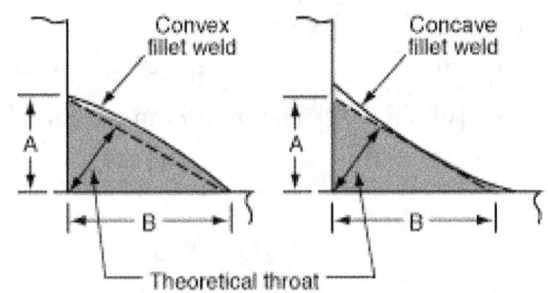

Unequal leg fillet weld

For unequal-leg fillet welds, the leg
lengths of the largest right triangle
which can be inscribed within the
fillet weld cross section. Size of weld
is A and B.

Figure 4—Fillet Welds

e. [GWR 103] Unless otherwise specified on the engineering drawing, the maximum acceptable fillet size shall be the size specified plus 50 percent or 4.8 mm (0.188 in), whichever is less, as permitted in section 4.11.

f. [GWR 104] The minimum acceptable actual throat shall equal or exceed the theoretical throat (reference figure 4).

4.8.4 Weldment Straightening

a. *Welds and adjacent base metal which have been deformed by the welding operation may be straightened.* [GWR 105] Prior to implementation, however, verification by NDE, destructive testing, and metallurgical evaluation that the process used for straightening does not degrade the weld or surrounding material below the specified design requirements shall be performed.

b. [GWR 106] Following weldment straightening, the weld and adjacent base metal shall be inspected in accordance with section 4.8.1.

c. [GWR 107] Weldments in which defects caused by weldment straightening are revealed shall not be acceptable.

4.8.5 Post-Weld Heat Treat Requirements

a. [GWR 108] Weldments that are subject to heat treatment operations shall be subsequently inspected to the surface quality requirements of the engineering drawing.

b. [GWR 109] Any required post-weld heat treatment processing shall be specified in the WPS.

4.9 Weld Joint Strength Requirements

4.9.1 Butt Joints

a. [GWR 110] If not otherwise specified in the design requirements, weld strength shall meet or exceed that of the parent material.

b. [GWR 111] Qualified WPSs shall be established to demonstrate the weld meets the strength required by design.

4.9.2 Fillet Welds

a. [GWR 112] Unless otherwise directed by the procuring agency, fillet weld shear strength shall meet or exceed 60 percent of the minimum ultimate tensile requirements of the weld.

b. [GWR 113] For fillet weld joints involving materials of different thicknesses having different ultimate tensile strength values, the minimum requirement for the shear joint shall be 60 percent of the lower of the minimum ultimate tensile requirements.

4.10 Weldment Quality Requirements

a. [GWR 114] Weldment quality requirements shall be established to ensure the weld meets design requirements for strength and integrity.

b. [GWR 115] The compliance of the weld with quality requirements shall be verified by mechanical testing.

4.11 Repair Welding

4.11.1 Allowable Repair Welding

a. [GWR 116] Additional welding operations shall be permitted to correct any unacceptable condition established per section 4.10, provided the repair welding parameters and procedures are specified in a qualified repair WPS, and the repair is contained within the original weld zone.

b. [GWR 117] Complete records of the repair welding operation, including identification of the repaired weldment, type of defect, and location of the repair weld, shall be retained in permanent records.

Examples of typical weld defects requiring repair are listed below.

> *(1) Undercut.*
> *(2) Lack of fill.*
> *(3) Concave root surface.*
> *(4) Incomplete joint penetration.*
> *(5) Crack and crack-like defects.*
> *(6) Oxides and porosity.*
> *(7) Lack of fusion.*

4.11.1.1 [GWR 118] No more than two weld repair attempts shall be made without approval of the MRB.

4.11.2 Repair Welding Requiring Disposition

At a minimum, the following conditions require MRB disposition by the procuring agency:

 a. *When more than two weld repair attempts have been performed at any one location.*

 b. *When the incorrect filler metal has been used.*

 c. *When a weldment has been post-weld heat treated to increase its strength and cannot be returned to the original drawing requirements with additional heat treatments following reweld.*

 d. *When finish machining has been completed prior to rewelding.*

 e. *When the repair extends outside the original weld zone.*

 f. *All FSW repairs.*

 g. *All friction plug repairs.*

4.11.3 Repair Welding Reinspection

[GWR 119] Reinspection of all repair weld areas shall be performed using the same methods/requirements as the original weld.

4.12 Quality Assurance

The supplier is responsible for the performance of all inspection requirements as specified herein.

 a. [GWR 120] Except as otherwise specified, the supplier shall use inspection facilities.

 b. [GWR 121] Except as otherwise specified, services approved by the procuring agency shall be used by the supplier.

 c. [GWR 122] Inspection and test records shall be kept complete and, upon request, made available to the procuring agency or its designated representative. *The procuring agency or its designated representative reserves the right to perform any or all of the inspections set forth in the specification to ensure that the end item conforms to the prescribed requirements.*

 d. [GWR 123] NDE procedures to be employed in inspection for weldment internal and surface quality requirements shall be qualified/validated as being capable of detecting the weldment quality criteria prescribed prior to inspection of the first production weld.

e. [GWR 124] The documented proof of capability shall be retained as a permanent record.

f. [GWR 125] Personnel performing visual weld inspections shall be certified in accordance with AWS QC1, Standard for AWS Certification of Welding Inspectors, or an equivalent standard as determined by the Quality Authority.

g. [GWR 126] Personnel performing NDE weld inspections shall be certified in accordance with NAS 410, NASA Certification and Qualification of Nondestructive Test Personnel, or an equivalent standard as determined by the Quality Authority.

4.12.1 Pre-Weld and Weld Inspection

a. [GWR 127] Documentation relative to the production weld shall be reviewed for conformance to section 4.

b. [GWR 128] Each production weld shall be certified that it was made within the range of operating parameters established in the WPS. *The contractor has the responsibility for certification.*

c. [GWR 129] Any deviations from operating parameters established in WPS shall be noted and referred to the procuring agency for disposition.

4.12.2 Post-Weld Inspection

4.12.2.1 Visual Inspection

a. [GWR 130] Each completed weldment, including the base metal, shall be inspected to ensure compliance with the requirements of sections 4.8.2, 4.8.3, 4.10, and as dictated by the class of weld for a minimum of 12.5 mm (0.5 in) on either side of the weld interface.

b. [GWR 131] The weld shall be in the as-welded condition for the initial inspection, with surface smut and loose oxide removed using a technique that does not smear metal or change the quality of the weld.

c. [GWR 132] Titanium weld deposit and heat-affected zone shall adhere to the color requirements of the accompanying detailed weld process specification identified in section 1.2.

4.12.2.2 Dimensional Inspection

[GWR 133] Dimensional inspection shall be performed on weldments to ensure compliance with the requirements of the design drawing for all weld classes.

4.12.2.3 Internal Quality Inspection

a. [GWR 134] NDE shall be performed to ensure compliance with the internal quality requirements of the design drawing established per section 4.10 for Class A, B, and C welds as noted in figure 5, Minimum Inspection Requirements.

b. [GWR 135] NDE procedures and techniques shall be qualified. *When a critical flaw size is specified, qualification includes verification of detectability of the critical size either by demonstration or by reference to the "standard" NDE methods and procedures identified in NASA-STD-5009, Nondestructive Evaluation Requirements for Fracture Critical Metallic Components. When reliability of inspection and critical flaw detection so dictate, redundant and/or complementing inspection techniques and procedures shall be employed.*

Method of Inspection	Weld Class		
	A	B	C
Visual	X	X	X
Dimensional	X	X	X
Surface	X	X	O
Volumetric	X	see note	O
Additional Inspection When Required by Drawing	X	X	X

Note: Class B welds shall be subjected to volumetric inspection if required by engineering design and specified by drawing or special instruction.

Figure 5—Minimum Inspection Requirements

4.12.2.4 Surface Quality Inspection

a. [GWR 136] NDE shall be performed to ensure compliance with the surface quality requirements of the design drawing for Class A and Class B welds.

b. [GWR 137] NDE procedures shall be qualified. *When a critical flaw size is specified, qualification includes verification of detectability of the critical size either by demonstration or by reference to the "standard" NDE methods and procedures identified in NASA-STD-5009.*

c. [GWR 138] When reliability of inspection and critical flaw detection so dictate, redundant and/or complementing inspection techniques and procedures shall be employed.

d. [GWR 139] Machined, ground, or otherwise mechanically worked weldments that have been subject to smearing of the weld material surface shall be etched to remove the masking material prior to penetrant application.

4.12.2.5 Records

a. [GWR 140] A continuous audit of weldment production quality shall be maintained.

b. [GWR 141] Weldment production quality audit records shall include, but not be limited to, the location of repairs, type of defects repaired, procedures used, and inches of repair per total inches of weld.

c. [GWR 142] Audit records of weldment production quality shall be made available to the procuring agency upon request.

APPENDIX A

GUIDANCE

A.1 Purpose

The purpose of this Appendix is to provide guidance, which is made available in the reference documents listed below. The latest edition of the referenced documents apply.

A.2 Reference Documents

A.2.1 Government Documents

Department of Defense

MIL-A-18455	Argon, Technical
MIL-PRF-27401	Propellant Pressurizing Agent, Nitrogen
MIL-PRF-27407	Propellant Pressurizing Agent, Helium
MIL-STD-1537	Electrical Conductivity Test for Verification of Heat Treatment of Aluminum Alloys, Eddy Current Method
MIL-STD-2154 (Cancelled)	Inspection, Ultrasonic, Wrought Metals, Process For
MIL-H-81200 (Cancelled)	Heat Treatment of Titanium and Titanium Alloys
MIL-HDBK-1823	Nondestructive Evaluation System Reliability Assessment
MIL-STD-2219A (Cancelled)	Fusion Welding For Aerospace Applications

Federal (authorized by the General Services Administration)

BB-C-101	Carbon Dioxide (CO_2): Technical and USP
BB-H-886	Hydrogen
BB-O-925A (Cancelled)	Oxygen, Technical, Gas and Liquid

National Aeronautics and Space Administration

NASA-STD-5001	Structural Design and Test Factors of Safety for Spaceflight Hardware
NASA-STD-5009	Nondestructive Evaluation Requirements for Fracture Critical Metallic Components
NASA-STD-5019	Fracture Control Requirements for Spaceflight Hardware
NASA-STD- 6016	Standard Materials and Processes Requirements for Spacecraft
NASA-STD-8719.17	NASA Requirements for Ground-Based Pressure Vessels and Pressurized Systems (PVS)

Johnson Space Center (JSC)

PRC-0001	Process Specification for the Manual Arc Welding of Aluminum Alloy Hardware
PRD-0002	Process Specification for the Manual Arc Welding of Titanium Alloy Hardware
PRC-0005	Process Specification for the Manual Arc Welding of Carbon Steel and Nickel Alloy Hardware
PRC-0008	Process Specification for the Qualification of Manual Arc Welders
PRC-0009	Process Specification for the Resistance Spot Welding of Battery and Electronic Assemblies
PRC-0010	Automatic and Machine Arc Welding of Steel and Nickel Alloy Hardware

Marshall Space Flight Center (MSFC)

MPR 8715.1	Marshall Safety, Health, and Environmental (SHE) Program
MSFC-SPEC-3679	Process Specification-Welding Aerospace Flight Hardware

A.2.2 Non-Government Documents

Aerospace Industries Association (AIA)/National Aerospace Standard (NAS)

AIA/NAS NAS976 (Inactive for New Design)	Electron-Beam Welding Machine—High Vacuum
AIA/NAS NAS1514	Radiographic Standard for Classification of Fusion Weld Discontinuities

American Society of Mechanical Engineers (ASME)

ASME B46.1	Surface Texture (Surface Roughness, Waviness, and Lay)

American Society for Testing Materials (ASTM)

ASTM E8/E8M	Standard Test Methods for Tension Testing of Metallic Materials
ASTM E2375	Standard Practice for Ultrasonic Testing of Wrought Products

American Welding Society (AWS)

AWS A2.4	Standard Symbols for Welding, Brazing, and Nondestructive Examination
AWS A3.0M/A3.0	Standard Welding Terms and Definitions Including Terms for Adhesive Bonding, Brazing, Soldering, Thermal Cutting, and Thermal Spraying
AWS A5.01M/A5.01	Welding Consumables—Procurement of Filler Metals and Fluxes
AWS A5.36/A5.36M	Specification for Carbon and Low-Alloy Steel Flux Cored Electrodes for Flux Cored Arc Welding and Metal Cored Electrodes for Gas Metal Arc Welding
AWS B4.0	Standard Methods for Mechanical Testing of Welds
AWS C6.1	Recommended Practices for Friction Welding
AWS C7.4/C7.4M	Process Specification and Operator Qualification for Laser Beam Welding
AWS D17.1/D17.1M	Specification for Fusion Welding for Aerospace Applications

AWS D17.2/D17.2M	Specification for Resistance Welding for Aerospace Applications
AWS D17.3/D17.3M	Specification for Friction Stir Welding of Aluminum Alloys for Aerospace Applications
AWS G2.4/G2.4M	Guide for the Fusion Welding of Titanium and Titanium Alloys

SAE International/American Material Specification (AMS)

SAE-AMS-2154	Inspection, Ultrasonic, Wrought Metals, Process For
SAE AMS2680	Electron-Beam Welding for Fatigue Critical Applications
SAE AMS 2770	Heat Treatment of Wrought Aluminum Alloy Parts
SAE AMS-H-81200	Heat Treatment of Titanium and Titanium Alloys

APPENDIX B

REQUIREMENTS COMPLIANCE MATRIX

B.1 Purpose

This Appendix provides a listing of requirements contained in this NASA Technical Standard for selection and verification of requirements by programs and projects. (*Note: Enter "Yes" to describe the requirement's applicability to the program or project; or enter "No" if the intent is to tailor, and enter how tailoring is to be applied in the "Rationale" column.*)

Section	Description	NASA-STD-5006A W/CHANGE 1 Requirement in this Standard	Applicable (Yes or No)	If No, Enter Rationale
1.3	Tailoring	[GWR 1] Tailoring of this NASA Technical Standard for application to a specific program or project shall be formally documented as part of program or project requirements and approved by the responsible Technical Authority in accordance with NPR 7120.5, NASA Space Flight Program and Project Management Requirements.		
2.1.1	Applicable Documents	[GWR 2] The latest issuances of cited documents shall apply unless specific versions are designated.		
2.1.2	Applicable Documents	[GWR 3] Non-use of specifically designated versions shall be approved by the responsible Technical Authority.		
2.4.2	Order of Precedence	[GWR 4] Conflicts between this NASA Technical Standard and other requirements documents shall be resolved by the responsible Technical Authority.		
		4. Requirements		
4.1	Specification of this NASA Technical Standard on Contracts	[GWR 5] When this NASA Technical Standard is specified on contract documents, a detailed weld process specification, as defined in NPR 7120.10, which meets the intent of this NASA Technical Standard shall be submitted.		
4.2	Joint Classes	[GWR 6] Welding performed using this NASA Technical Standard shall be classified in accordance with the service of the joints as follows in the next sections.		
4.2.1.1	Class A	[GWR 7] Class A welds shall require visual, dimensional, surface, and volumetric inspections, and additional inspection when required by engineering drawing.		
4.2.2.1	Class B	[GWR 8] Class B welds shall require visual, dimensional, and surface inspections, and additional inspection when required by engineering drawing.		
4.2.2.2	Class B	[GWR 9] Class B welds shall be subjected to volumetric inspection if required by engineering design and specified by drawing or special instruction.		

NASA-STD-5006A W/CHANGE 1

Section	Description	Requirement in this Standard	Applicable (Yes or No)	If No, Enter Rationale
		NASA-STD-5006A W/CHANGE 1		
4.2.2.3	Class B	[GWR 10] Weld requiring fail-safe capability shall be classified as a Class B joint.		
4.2.3.1	Class C	[GWR 11] Class C welds shall require visual and dimensional inspections, and, additional inspection when required by engineering drawing.		
4.2.3.2	Class C	[GWR 12] Class C joints shall require weld integrity verification based on function of the joint (e.g., seal welds require leak testing commensurate with the leak rate requirement).		
		4.3 Equipment		
4.3.1a	Welding Equipment	[GWR 13] Automatic, semiautomatic, manual, and machine welding shall be accomplished using equipment containing calibrated dials, meters, or recorders that quantitatively indicate process parameters.		
4.3.1b	Welding Equipment	[GWR 14] All joining equipment (including manual) shall be capable of producing joints that meet the requirements specified herein.		
4.3.1.1a	Acceptance Testing	[GWR 15] New, repaired, relocated, or modified welding machines and equipment for automatic and machine welding shall be acceptance-tested prior to processing of flight hardware.		
4.3.1.1b	Acceptance Testing	[GWR 16] Machines and equipment shall meet the requirements of the applicable purchase specification, design specification, or modification order.		
4.3.1.1c	Acceptance Testing	[GWR 17] Power supplies and supporting components (electrical or mechanical or both) shall be capable of operating reliably within the range of parameters and duty cycle to be used for joining production parts.		
4.3.1.2a	Calibration	[GWR 18] Measuring instruments, meters, gages, or direct reading electrical control circuits to be used for automatic, semiautomatic, and machine joining operations shall be initially calibrated and periodically recalibrated to maintain adequate performance.		
4.3.1.2b	Calibration	[GWR 19] Maintenance performed on measuring instruments, meters, gages, or direct reading electrical control circuits to be used for automatic, semiautomatic, and machine performance joining operations shall require recalibration to maintain adequate performance.		
4.3.1.2c	Calibration	[GWR 20] Measuring instruments, meters, gages, or direct reading electrical control circuits to be used for automatic, semiautomatic, and machine joining operations shall be initially calibrated and periodically recalibrated to maintain adequate performance or when any maintenance is performed that may have changed calibration.		
4.3.1.3a	Maintenance and Records	[GWR 21] Welding machines shall be provided with adequate periodic maintenance service so that acceptable welds can be produced using qualified welding procedure specifications.		

NASA-STD-5006A W/CHANGE 1

Section	Description	Requirement in this Standard	Applicable (Yes or No)	If No, Enter Rationale
		NASA-STD-5006A W/CHANGE 1		
4.3.1.3b	Maintenance and Records	[GWR 22] A current record of each maintenance repair or functional check shall be maintained for each welding machine.		
4.3.2a	Tooling and Fixtures	[GWR 23] Tooling and fixtures used in the joining operation shall be constructed of nonmagnetic materials that do not affect the welding arc or beam, or that are not detrimental to the weld quality.		
4.3.2b	Tooling and Fixtures	[GWR 24] Tooling and fixtures shall not be a source of contamination of the joint.		
4.3.2c	Tooling and Fixtures	[GWR 25] Magnetic materials, when used, shall be degaussed prior to welding.		
4.3.2d	Tooling and Fixtures	[GWR 26] Degaussing, when necessary, shall be controlled by the WPS for the successful completion of the weld.		
4.3.2e	Tooling and Fixtures	[GWR 27] Tooling and fixtures required to ensure compliance with dimensional requirements of section 4.8.3 shall be identified on the WPS.		
		4.4 Materials		
4.4.1a	Base Metals	[GWR 28] Unless otherwise specified or approved by the procuring agency, base metal alloy shall conform to applicable government and/or industry specifications for each alloy group.		
4.4.1b	Base Metals	[GWR 29] The base metal, material condition, and appropriate specification shall be recorded in the WPS.		
4.4.1c	Base Metals	[GWR 30] Weld start and runoff tabs, when used, shall be of the same alloy as the material being joined and be welded with the same filler metal specified on the drawing or WPS.		
4.4.2a	Filler Metals	[GWR 31] Unless otherwise specified or approved by the procuring agency, filler metal alloy shall conform to applicable government and/or industry specifications for each alloy group.		
4.4.2b	Filler Metals	[GWR 32] Specifications used to procure filler metals shall include provisions to mitigate the possibility of two different filler wires being errantly mixed together on a single spool or in a filler rod container.		
4.4.2c	Filler Metals	[GWR 33] Weld filler metals and the appropriate specifications shall be recorded in the WPS.		
4.4.3a	Shielding Gas	[GWR 34] Welding-grade gases conforming to the applicable industry or military specifications shall be used for gas shielding.		
4.4.3b	Shielding Gas	[GWR 35] The shield gas type, specification, and flow rate shall be recorded in the WPS.		
4.4.4a	Tungsten Electrodes	[GWR 36] Tungsten electrodes shall conform to the applicable industry or military specifications.		

NASA-STD-5006A W/CHANGE 1

Section	Description	NASA-STD-5006A W/CHANGE 1 Requirement in this Standard	Applicable (Yes or No)	If No, Enter Rationale
4.4.4b	Tungsten Electrodes	[GWR 37] The electrode diameter, electrode tip shape, alloy composition, and specification shall be recorded as a part of the WPS.		
4.4.5.1	Friction Stir Welding Pin Tools	[GWR 38] Pin and shoulder service life shall be demonstrated to meet the intended use.		
4.4.5.2	Friction Stir Welding Pin Tools	[GWR 39] Pins and shoulders shall be limited to the demonstrated life.		
4.4.5.3	Friction Stir Welding Pin Tools	[GWR 40] Pin tool design, materials, and service life shall be recorded in the WPS.		
4.4.5.3a	Friction Stir Welding Pin Tools	[GWR 41] Pins and shoulders that have reached the specified service life shall be marked and removed from service to preclude an accidental future use in the FSW production process.		
4.4.5.3b	Friction Stir Welding Pin Tools	[GWR 42] If used for more than one weld joint, pins and shoulders shall be cleaned and inspected before reuse on production hardware.		
		4.5 Weld Procedure and Performance Qualification		
4.5.1a	Welder Performance Qualification	[GWR 43] Operators of automatic, semiautomatic, machine, or manual welding equipment shall be certified for the applicable process.		
4.5.1b	Welder Performance Qualification	[GWR 44] Suppliers shall define the weld certification process in their detailed weld process specification.		
4.5.2a	Welding Procedure Specification	[GWR 45] Prior to first production of parts, or when changes are made to essential variables of the WPS, qualification joints shall be made to establish a satisfactory WPS for each different configuration of A, B, and C classes of welds.		
4.5.2b	Welding Procedure Specification	[GWR 46] Variables considered essential shall be so identified in the WPS.		
4.5.2.1a	Classes A and B Joints	[GWR 47] Classes A and B joints shall be qualified with joints that simulate the production part with respect to section thickness, alloy, heat-treat condition, joint preparation, pre-weld cleaning, fit-up, position, and post-weld operations.		
4.5.2.1b	Classes A and B Joints	[GWR 48] The joints shall be processed in either the actual production fixture or in a test fixture simulating the production fixture using the production welding equipment.		
4.5.2.1c	Classes A and B Joints	[GWR 49] Base metal for qualification joining tests shall be identified by lot or heat number, type, and condition and maintain identification through all evaluation processes.		
4.5.2.1d	Classes A and B Joints	[GWR 50] The qualification weld shall be subjected to metallurgical evaluation and the same post-weld inspections and processes as the production parts, including reinforcement removal, mechanical deformation, stress relief, and thermal treatments associated with artificial aging or any operation affecting mechanical properties.		

NASA-STD-5006A W/CHANGE 1

Section	Description	NASA-STD-5006A W/CHANGE 1 Requirement in this Standard	Applicable (Yes or No)	If No, Enter Rationale
4.5.2.2a	Joining Parameters	[GWR 51] As a minimum, all essential joining variables (such as voltage, current, rate of travel, position, and filler-wire feed rate) shall be recorded during qualification welding.		
4.5.2.2b	Joining Parameters	[GWR 52] Manual weld parameters and operating parameter ranges shall be established during the WPS qualification.		
4.5.2.2c	Joining Parameters	[GWR 53] The WPS shall document all pre-welding operations, setup conditions, welding equipment, and any pertinent information about the welding system used which affects the joining operation.		
4.5.2.3a	Parameter Tolerances	[GWR 54] For automatic, semiautomatic, and machine joining, parameter tolerances may be used and shall be listed in the qualified WPS.		
4.5.2.3b	Parameter Tolerances	[GWR 55] Test samples representing the minimum and maximum heat input shall be processed to verify acceptable welds and the results documented in the PQR.		
4.5.3	Welding Procedure Specification Qualification	[GWR 56] All test and inspection results used to verify the weld integrity shall be recorded on the PQR.		
4.5.4.1	Records	[GWR 57] Records of test specimens that meet the acceptance requirements of this process specification shall be signed and dated by a Quality Assurance (QA) representative as an accurate record of the welding and testing of the procedure test weldment.		
4.5.4.2	Records	[GWR 58] The WPS and PQR shall be prepared and retained as long-term temporary records in accordance with NPR 1441.1, NASA Records Retention Schedules, with the current WPS being maintained at the welding station.		
4.5.4.3	Records	[GWR 59] All WPSs and PQRs shall be maintained and made available for review by the responsible NASA Engineering Authority before production of hardware covered under this NASA Technical Standard.		
		4.6 Pre-Weld Operations		
4.6.1	Joint Design	[GWR 60] All joints shall be documented on a WPS, design drawing, or other suitable document.		
4.6.2a	Pre-Weld Cleaning	[GWR 61] Pre-weld cleaning of filler materials and surfaces to be welded to remove contaminants that are detrimental to weld quality shall be accomplished in an environment that will not degrade the quality of the weld.		
4.6.2b	Pre-Weld Cleaning	[GWR 62] Cleanliness shall be maintained during welding.		
4.6.2c	Pre-Weld Cleaning	[GWR 63] Pre-weld and interpass cleaning requirements shall be included in the WPS.		
		4.7 Production Welding		
4.7.1a	Equipment Operational Check	[GWR 64] A welding equipment operational readiness check shall be made immediately prior to a production weld to verify the equipment is operating properly.		

NASA-STD-5006A W/CHANGE 1

	NASA-STD-5006A W/CHANGE 1			
Section	Description	Requirement in this Standard	Applicable (Yes or No)	If No, Enter Rationale
4.7.1b	Equipment Operational Check	[GWR 65] The equipment operational readiness check criteria shall be provided to the procuring agency.		
4.7.2a	Temperature Control	[GWR 66] Pre-heat, interpass, and post-heat temperatures shall be controlled so as not to degrade the properties of the material being welded.		
4.7.2b	Temperature Control	[GWR 67] The parameters of pre-heat, interpass, and post-heat temperatures shall be recorded in an applicable WPS.		
4.7.3a	Tack Welding	[GWR 68] Tack welding is allowable and shall either be removed or become a part of the finished weld (i.e., tack welds are to be completely consumed by the final weldment).		
4.7.3b	Tack Welding	[GWR 69] Tack welds that become part of the finished weld shall be performed by certified welders in accordance with certified procedures meeting the requirements of this NASA Technical Standard.		
4.7.3c	Tack Welding	[GWR 70] After the weldment is completed, the tack areas shall be evaluated to the requirements of the finished weld.		
4.7.3d	Tack Welding	[GWR 71] Tack welding requirements shall be included in the WPS.		
4.7.4.1a	Classes A and B Joints	[GWR 72] The technique of welding the initial passes from both sides where the weld roots overlap beneath the exposed surfaces (reference figure 2 (A), Welding Techniques) shall be permitted only if the root of the first pass is removed to sound metal prior to placement of the first weld pass from the second side.		
4.7.4.1b	Classes A and B Joints	[GWR 73] Joints which have prepared grooves from one or both sides (reference figure 2 (B) and (C)) and/or multi-pass welds shall have a weld land that is completely penetrated on the initial pass.		
4.7.4.1c	Classes A and B Joints	[GWR 74] Adequate nondestructive evaluation (NDE) procedures shall be employed to ensure that the weld root has been exposed by machining.		
4.7.4.1d	Classes A and B Joints	[GWR 75] All penetration weld passes shall have no visual evidence of improper fusion or presence of dross.		
4.7.4.1e	Classes A and B Joints	[GWR 76] Square groove welds shall be completely penetrated from one side (reference figure 2 (D)).		
4.7.4.2a	Class C Joints	[GWR 77] The technique of welding and joint geometry shall be as stated on the engineering drawing and the WPS.		
4.7.4.2	Class C Joints	[GWR 78] Any deviation regarding weld technique and joint geometry shall be approved by the procuring agent prior to use.		
4.7.4.2c	Class C Joints	[GWR 79] Partial penetration groove welds shall be used only for Class C joints.		
4.7.5	Welding Procedure	[GWR 80] Production welding shall be accomplished according to a qualified WPS.		

APPROVED FOR PUBLIC RELEASE — DISTRIBUTION IS UNLIMITED

NASA-STD-5006A W/CHANGE 1

Section	Description	Requirement in this Standard	Applicable (Yes or No)	If No, Enter Rationale
		NASA-STD-5006A W/CHANGE 1		
4.7.6a	Procedure Departure	[GWR 81] Departure from the qualified WPS during production welding shall require withholding the part for MRB disposition.		
4.7.6b	Procedure Departure	[GWR 82] The cause for departure shall be determined.		
4.7.6c	Procedure Departure	[GWR 83] Corrective action shall be taken prior to further production welding.		
		4.8 Post-Weld Operations		
4.8.1	Inspection	[GWR 84] Each completed weldment, including the base metal, shall be inspected to ensure compliance with the requirements of sections 4.8.2, 4.8.3, and 4.10, and as dictated by the class of weld for a minimum of 12.5 mm (0.5 in) on either side of the weld interface.		
4.8.2a	General Visual/Surface Requirements	[GWR 85] Weld deposits, buildup, and root reinforcement shall comply with the criteria outlined in the accompanying detailed weld process specification that is submitted in support of this NASA Technical Standard.		
4.8.2b	General Visual/Surface Requirements	[GWR 86] The edge of the weld deposit shall blend into the base metal without unfused overlaps or undercut.		
4.8.2c	General Visual/Surface Requirements	[GWR 87] Weld face and root sides shall be free of surface cracks, crater cracks, and other defects open to the surface.		
4.8.2d	General Visual/Surface Requirements	[GWR 88] Weld deposits shall be free of open voids or unfused overlapping folds or other lack of fusion.		
4.8.2e	General Visual/Surface Requirements	[GWR 89] Undercutting, concavity, lack of fill, or root concavity shall be unacceptable in any weld where it occurs as a sharp notch or where the depth reduces the material thickness below the minimum thickness specified on the applicable drawing.		
		4.8.3 Dimensional Requirements		
4.8.3.1	Mismatch	[GWR 90] If not specifically addressed in drawing tolerances or by specified welding standards, allowable post-weld mismatch shall be governed by overall drawing tolerances.		
4.8.3.2a	Peaking	[GWR 91] If not specifically addressed in drawing tolerances or by specified welding standards, allowable post-weld peaking of the welded joint and adjacent base metal shall be governed by overall drawing tolerances.		
4.8.3.2b	Peaking	[GWR 92] A standard template or other device having specified reference points shall be used for determination of peaking.		
4.8.3.4a	Weld Reinforcement Removal	[GWR 93] Weld bead reinforcement removal shall not thin the weld or parent metal below drawing dimensional requirements.		
4.8.3.4b	Weld Reinforcement Removal	[GWR 94] When flush contour is required by the welding symbol, weld reinforcement shall not exceed 0.4 mm (0.015 in).		

NASA-STD-5006A W/CHANGE 1

NASA-STD-5006A W/CHANGE 1

Section	Description	Requirement in this Standard	Applicable (Yes or No)	If No, Enter Rationale
4.8.3.4c	Weld Reinforcement Removal	[GWR 95] Metal removal shall be such that the reworked area blends smoothly (e.g., 3.175 mm (0.125 in) radius) with adjacent material without abrupt sectional changes.		
4.8.3.4d	Weld Reinforcement Removal	[GWR 96] Weldments that are machined ground or otherwise mechanically worked causing disruption or smearing of the material surface shall be etched to remove the masking material before penetrant application.		
4.8.3.5.1a	Fillet Welds	[GWR 97] Fillet weld fusion of the root shall have a minimum of 10 percent penetration of base metal thickness of the thinnest member of the root of the joint as determined by evaluation of transverse sections taken from the qualification welds.		
4.8.3.5.1a(1)	Fillet Welds	[GWR 98] The minimum penetration shall be verified by destructive test/metallography.		
4.8.3.5.1a(2)	Fillet Welds	[GWR 99] Weld parameters used to successfully and repeatedly complete the fillet welds shall be entered into the WPS and used for actual welding.		
4.8.3.5.1b	Fillet Welds	[GWR 100] Fillet weld fusion of the root (reference figure 4, Fillet Welds) shall be determined by evaluation of transverse sections taken from the qualification welds.		
4.8.3.5.1c	Fillet Welds	[GWR 101] Intermittent fillet welds shall have fusion of the root throughout the specified length. Unless otherwise specified on the engineering drawing, the fillet may be extended by 6.35 mm (0.25 in) at each end without penetration in the extension.		
4.8.3.5.1d	Fillet Welds	[GWR 102] The minimum acceptable fillet size shall be that specified by the engineering drawing.		
4.8.3.5.1e	Fillet Welds	[GWR 103] Unless otherwise specified on the engineering drawing, the maximum acceptable fillet size shall be the size specified plus 50 percent or 4.8 mm (0.188 in), whichever is less, as permitted in section 4.11.		
4.8.3.5.1f	Fillet Welds	[GWR 104] The minimum acceptable actual throat shall equal or exceed the theoretical throat (reference figure 4).		
4.8.4a	Weldment Straightening	*Welds and adjacent base metal which have been deformed by the welding operation may be straightened.* [GWR 105] Prior to implementation, however, verification by NDE, destructive testing, and metallurgical evaluation that the process used for straightening does not degrade the weld or surrounding material below the specified design requirements shall be performed.		
4.8.4b	Weldment Straightening	[GWR 106] Following weldment straightening, the weld and adjacent base metal shall be inspected in accordance with section 4.8.1.		
4.8.4c	Weldment Straightening	[GWR 107] Weldments in which defects caused by weldment straightening are revealed shall not be acceptable.		

APPROVED FOR PUBLIC RELEASE — DISTRIBUTION IS UNLIMITED

NASA-STD-5006A W/CHANGE 1

Section	Description	Requirement in this Standard	Applicable (Yes or No)	If No, Enter Rationale
		NASA-STD-5006A W/CHANGE 1		
4.8.5a	Post-Weld Heat Treat Requirements	[GWR 108] Weldments that are subject to heat treatment operations shall be subsequently inspected to the surface quality requirements of the engineering drawing.		
4.8.5b	Post-Weld Heat Treat Requirements	[GWR 109] Any required post-weld heat treatment processing shall be specified in the WPS.		
		4.9 Weld Joint Strength Requirements		
4.9.1a	Butt Joints	[GWR 110] If not otherwise specified in the design requirements, weld strength shall meet or exceed that of the parent material.		
4.9.1b	Butt Joints	[GWR 111] Qualified WPSs shall be established to demonstrate the weld meets the strength required by design.		
4.9.2a	Fillet Welds	[GWR 112] Unless otherwise directed by the procuring agency, fillet weld shear strength shall meet or exceed 60 percent of the minimum ultimate tensile requirements of the weld.		
4.9.2b	Fillet Welds	[GWR 113] For fillet weld joints involving materials of different thicknesses having different ultimate tensile strength values, the minimum requirement for the shear joint shall be 60 percent of the lower of the minimum ultimate tensile requirements.		
4.10a	Weldment Quality Requirements	[GWR 114] Weldment quality requirements shall be established to ensure the weld meets design requirements for strength and integrity.		
4.10b	Weldment Quality Requirements	[GWR 115] The compliance of the weld with quality requirements shall be verified by mechanical testing.		
		4.11 Repair Welding		
4.11.1a	Allowable Repair Welding	[GWR 116] Additional welding operations shall be permitted to correct any unacceptable condition established per section 4.10, provided the repair welding parameters and procedures are specified in a qualified repair WPS, and the repair is contained within the original weld zone.		
4.11.1b	Allowable Repair Welding	[GWR 117] Complete records of the repair welding operation, including identification of the repaired weldment, type of defect, and location of the repair weld, shall be retained in permanent records.		
4.11.1.1	Allowable Repair Welding	[GWR 118] No more than two weld repair attempts shall be made without approval of the MRB.		
4.11.3	Repair Welding Reinspection	[GWR 119] Reinspection of all repair weld areas shall be performed using the same methods/requirements as the original weld.		
4.12a	Quality Assurance	[GWR 120] Except as otherwise specified, the supplier shall use inspection facilities.		
4.12b	Quality Assurance	[GWR 121] Except as otherwise specified, services approved by the procuring agency shall be used by the supplier.		

APPROVED FOR PUBLIC RELEASE — DISTRIBUTION IS UNLIMITED

NASA-STD-5006A W/CHANGE 1

Section	Description	Requirement in this Standard	Applicable (Yes or No)	If No, Enter Rationale
		NASA-STD-5006A W/CHANGE 1		
4.12c	Quality Assurance	[GWR 122] Inspection and test records shall be kept complete and, upon request, made available to the procuring agency or its designated representative.		
4.12d	Quality Assurance	[GWR 123] NDE procedures to be employed in inspection for weldment internal and surface quality requirements shall be qualified/validated as being capable of detecting the weldment quality criteria prescribed prior to inspection of the first production weld.		
4.12e	Quality Assurance	[GWR 124] The documented proof of capability shall be retained as a permanent record.		
4.12f	Quality Assurance	[GWR 125] Personnel performing visual weld inspections shall be certified in accordance with AWS QC1, Standard for AWS Certification of Welding Inspectors, or an equivalent standard as determined by the Quality Authority.		
4.12g	Quality Assurance	[GWR 126] Personnel performing NDE weld inspections shall be certified in accordance with NAS 410, NASA Certification and Qualification of Nondestructive Test Personnel, or an equivalent standard as determined by the Quality Authority.		
4.12.1a	Pre-Weld and Weld Inspection	[GWR 127] Documentation relative to the production weld shall be reviewed for conformance to section 4.		
4.12.1b	Pre-Weld and Weld Inspection	[GWR 128] Each production weld shall be certified that it was made within the range of operating parameters established in the WPS. The contractor has the responsibility for certification.		
4.12.1c	Pre-Weld and Weld Inspection	[GWR 129] Any deviations from operating parameters established in WPS shall be noted and referred to the procuring agency for disposition.		
		4.12.2 Post Weld Inspection		
4.12.2.1a	Visual Inspection	[GWR 130] Each completed weldment, including the base metal, shall be inspected to ensure compliance with the requirements of sections 4.8.2, 4.8.3, 4.10, and as dictated by the class of weld for a minimum of 12.5 mm (0.5 in) on either side of the weld interface.		
4.12.2.1b	Visual Inspection	[GWR 131] The weld shall be in the as-welded condition for the initial inspection, with surface smut and loose oxide removed using a technique that does not smear metal or change the quality of the weld.		
4.12.2.1c	Visual Inspection	[GWR 132] Titanium weld deposit and heat-affected zone shall adhere to the color requirements of the accompanying detailed weld process specification identified in section 1.2.		
4.12.2.2	Dimensional Inspection	[GWR 133] Dimensional inspection shall be performed on weldments to ensure compliance with the requirements of the design drawing for all weld classes.		

NASA-STD-5006A W/CHANGE 1

Section	Description	Requirement in this Standard	Applicable (Yes or No)	If No, Enter Rationale
4.12.2.3a	Internal Quality Inspection	[GWR 134] NDE shall be performed to ensure compliance with the internal quality requirements of the design drawing established per section 4.10 for Class A, B, and C welds as noted in figure 5, Minimum Inspection Requirements. **Figure 5—Minimum Inspection Requirements** *Note: Class B welds shall be subjected to volumetric inspection if required by engineering design and specified by drawing or special instruction.*		
4.12.2.3b	Internal Quality Inspection	[GWR 135] NDE procedures and techniques shall be qualified.		
4.12.2.4a	Surface Quality Inspection	[GWR 136] NDE shall be performed to ensure compliance with the surface quality requirements of the design drawing for Class A and Class B welds.		
4.12.2.4b	Surface Quality Inspection	[GWR 137] NDE procedures shall be qualified. When a critical flaw size is specified, qualification includes verification of detectability of the critical size either by demonstration or by reference to the "standard" NDE methods and procedures identified in NASA-STD-5009.		
4.12.2.4c	Surface Quality Inspection	[GWR 138] When reliability of inspection and critical flaw detection so dictate, redundant and/or complementing inspection techniques and procedures shall be employed.		
4.12.2.4d	Surface Quality Inspection	[GWR 139] Machined, ground, or otherwise mechanically worked weldments that have been subject to smearing of the weld material surface shall be etched to remove the masking material prior to penetrant application.		
4.12.2.5a	Records	[GWR 140] A continuous audit of weldment production quality shall be maintained.		
4.12.2.5b	Records	[GWR 141] Weldment production quality audit records shall include, but not be limited to, the location of repairs, type of defects repaired, procedures used, and inches of repair per total inches of weld.		
4.12.2.5c	Records	[GWR 142] Audit records of weldment production quality shall be made available to the procuring agency upon request.		

Figure 5 table embedded in section 4.12.2.3a:

Method of Inspection	Weld Class		
	A	B	C
Visual	X	X	X
Dimensional	X	X	X
Surface	X	X	O
Volumetric	X	see note	O
Additional Inspection When Required by Drawing	X	X	X